The Thinking Tree

HOMESTEADING POETRY

50 POEMS & PROMPTS

CREATIVE LANGUAGE ARTS
WRITING, SPELLING, READING, COPYWORK & ART ACTIVITIES

FUN-SCHOOLING
WITH THINKING TREE BOOKS

Curriculum Design & Illustrations
by Sarah Janisse Brown

This book is set in the Dyslexie Font,
designed by Christian Boer

Copyright © 2025

Published by:

The Thinking Tree , LLC

Phone: 317.622.8852

Email: Contact@FunSchooling.com Website:

FunSchooling.com

Name:

INTRODUCTION:

Welcome to Homesteading Poetry — a creative language arts journey through the rhythms of rural life, nature's beauty, and the joys of simple living. This book is filled with 50 illustrated poems and writing prompts designed to enrich your child's love for language while celebrating the traditions of homesteading, gardening, foraging, food preservation, and living close to the land.

Through poetry, copywork, spelling, and hands-on creative activities, children explore meaningful topics such as composting, canning, herbs, nature walks, and farm chores — all through the lens of gentle rhyme and reflective storytelling.

Whether you live in the country, the suburbs, or the city, this book is a doorway to deeper connection with the natural world and the skills that keep it thriving.

PARENT & TEACHER INSTRUCTIONS

This workbook is ideal for children ages 8 to 14, though younger learners can join in with help.

Each unit includes:

- An original homesteading-themed poem
- A fill-in-the-blank spelling or vocabulary practice version of the poem
- Copywork and handwriting pages
- Creative illustrations to color
- Art, nature, or writing prompts inspired by the theme

HOW TO USE THIS BOOK:

- Read the Poem Aloud Together—Reading the poem three times helps build fluency, comprehension, and rhythm awareness.

- Complete the Fill-in-the-Blank Page—This reinforces vocabulary and spelling while encouraging attention to detail.

- Copy the Poem by Hand— Use the lined pages provided to practice penmanship and internalize the poem's message. Encourage careful handwriting.

- Color the Illustration or Draw Your Own—Coloring engages both sides of the brain and allows students to make the poem visually personal.

- Expand with Writing or Art Prompts—Encourage students to write their own poems, journal about their homesteading experiences, or illustrate scenes from the poem.

OPTIONAL EXTENSIONS:

- Add nature walks, herb identification, or garden tasks to match the theme.

- Create a poetry binder or illustrated journal.

- Read aloud poetry by other nature writers or homesteaders.

- Use this book as a seasonal supplement to your homeschool or co-op language arts curriculum.

POETRY READING

Read the poem or story three times.

BREAD & BUTTER

by Deri Jansma

Making things
is fun to do—
bread and jam
and butter too.
Measuring, mixing,
always fun,
baking bread,
until it's done.
Churning butter,
cooking jam,
filling up
the jars and cans.
And when the making time must end,
I can't wait
To do it all again.

READING & COPYWORK

Copy the poem or story below and add color to the illustration.

_____

_____

_____

_____

_____

_____

_____

_____

_____

_____

_____

_____

_____

_____

_____

_____

_____

_____

_____

VANILLA SUGAR

POETRY READING

Read the poem or story three times.

CANNING

By Michael Merriman

Canning food was once a staple.

Now they call it an art, almost a fable.

Cleaning the jars and the lids,

making jams from berries and figs.

Fresh veggies from the garden-

better to can than let them go rotten.

From generation to generation,

we are blessed to celebrate this tradition.

CANNING

Canning food was once a ▢▢▢▢▢.

Now they call it an art, ▢▢▢▢▢▢ a fable.

▢▢▢▢▢▢▢ the jars and the lids,

making jams from ▢▢▢▢▢▢▢ and figs.

Fresh veggies from the ▢▢▢▢▢-

▢▢▢▢▢▢ to can than let them go rotten.

From ▢▢▢▢▢▢▢▢▢ to generation,

we are ▢▢▢▢▢▢ to celebrate this tradition.

READING & COPYWORK

Copy the poem or story below and add color to the illustration.

_____

_____

_____

_____

_____

_____

_____

_____

_____

_____

_____

_____

POETRY READING

Read the poem or story three times.

A CUP FULL OF SUNSHINE

by Dale Smart-Felt

We skip through the garden with baskets in hand,

picking herbs, leaves, and flowers just as we planned.

We don't take them all- we leave some to grow

for bees and the breeze and the seeds they might sow.

We hang them to dry on the sill and the line;

Each one chosen with care, our blend will be fine!

Its tea we're making, to sip and to share-

for grumps or for giggles, sickness or shiny hair.

There's a blend to help everything, never you fear;

we simmer it gently and pour it with cheer.

A warm, tasty hug when the cold days are near,

it's yummy, it's natural, it helps us feel bright-

a cup full of sunshine, morning or night!

Sage   Cornflower   Marshmallow   Burnet

READING & COPYWORK

Copy the poem or story below and add color to the illustration.

_____

_____

_____

_____

_____

_____

_____

_____

_____

_____

_____

_____

_____

_____

_____

_____

_____

_____

_____

_____

_____

_____

Celandine   Blue Vervain   Tansy   Hyssop

POETRY READING

Read the poem or story three times.

APPLE BUTTER

by Dixie Miller & Sarah J. Brown

Climb the tree,
search under leaves,
for ripe and sweet
fresh apples.

Pick enough for now and later.
take a bite, enjoy the flavor-
gather up a bushel
of fresh apples.

Spicy, tart, and sweet,
peeling, slicing, mashing,
Cooking, stirring, mixture bubbling-
Mouth-watering!

The smell is wonderful,
Jars and jars for all year long
Homemade apple butter.

READING & COPYWORK

Copy the poem or story below and add color to the illustration.

POETRY READING & WRITING

Read the poem or story and then fill in the missing words.

FABRIC, NEEDLE & THREAD

by MaKenzie Alberdi

One stitch, two stitch,

three stitches and more!

Making this fabric into

something it wasn't before.

Will it be a bag or a dress?

Maybe a blanket for the bed?

You can make so many things,

with fabric, needle, and thread!

FABRIC, NEEDLE & THREAD

One ☐☐☐☐, two stitch

three stitches and ☐☐☐☐!

Making this ☐☐☐☐☐ into

something it wasn't ☐☐☐☐☐.

Will it be a bag or a ☐☐☐☐☐?

Maybe a ☐☐☐☐☐ for the bed?

You can make so many ☐☐☐☐☐,

with fabric, ☐☐☐☐☐, and thread!

READING & COPYWORK

Copy the poem or story below and add color to the illustration.

_____

_____

_____

_____

_____

_____

_____

_____

_____

_____

_____

_____

_____

POETRY READING

Read the poem or story three times.

BEESWAX CANDLES

By Gideon Miller & Sarah J. Brown

Of all the things that bees can make,

some say honey takes the cake.

But I am certain of the fact

that brilliant light

comes with the wax!

Take the bee's wax,

heat it up

in a bowl or in a cup.

Colors, scents, and shapes- you pick!

Pour it carefully, hold the wick.

The candle is cooled, now it is night-

Homemade candle, shining bright!

READING & COPYWORK

Copy the poem or story below and add color to the illustration.

_____

_____

_____

_____

_____

_____

_____

_____

_____

_____

_____

_____

_____

_____

_____

_____

POETRY READING

Read the poem or story three times.

COMPOSTING

by Carrie Rice-Richardson

Beneath the earth where secrets lie,
wiggly worms come passing by.
They turn the scraps we love to throw,
to rich dark dirt seeds will use to grow.
From rot and decay life finds a start,
composting into a work of art.

COMPOSTING

☐☐☐☐☐☐ the earth where secrets lie,

wiggly worms come ☐☐☐☐☐☐ by.

They turn the ☐☐☐☐☐☐ we love to throw,

to rich dark dirt ☐☐☐☐☐ will use to grow.

From rot and ☐☐☐☐☐ life finds a start,

composting into a ☐☐☐☐ of art.

READING & COPYWORK

Copy the poem or story below and add color to the illustration.

_____

_____

_____

_____

_____

_____

_____

_____

_____

_____

_____

_____

_____

_____

_____

_____

_____

_____

POETRY READING

Read the poem or story three times.

GROWING OUR OWN FOOD

by Deri Jansma & Sarah Janisse Brown

We like to grow the food we eat-
fresh berries are such a healthy treat!
Lettuce, basil, carrots and cilantro,
celery, mint and potatoes!

Tending the soil and watering weekly
helps our food grow ever so sweetly.
Making sure there's enough sunlight,
double checking everything is just right.

We love to harvest the food we grew
And share it with our loved ones too!
With baskets full of greens and fruit,
We can't complain of muddy boots!

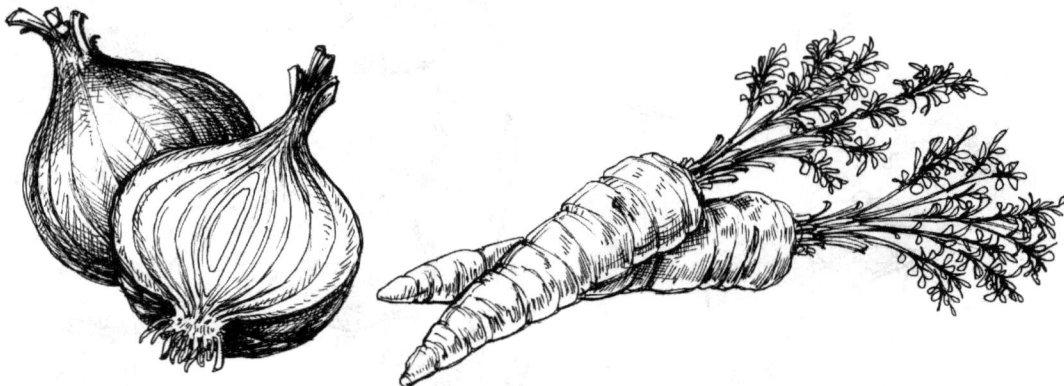

READING & COPYWORK

Copy the poem or story below and add color to the illustration.

_____

_____

_____

_____

_____

_____

_____

_____

_____

_____

_____

_____

_____

_____

_____

_____

_____

POETRY READING

Read the poem or story three times.

FREE FOOD!

by Bear Sutcliffe

In spring and summer, we spend our hours

foraging for herbs and flowers.

Three shapes of leaves on a sassafras tree

help me to know where the yummy root beer leaves will be.

Elderberry, blackberry, and wild strawberry for jelly,

wild mint for when I have an upset belly.

Wild green for a salad, mushrooms and ramps with our steak,

We give back to the earth for the food that we take!

FREE FOOD!

In spring and ⬜⬜⬜⬜⬜⬜⬜, we spend our hours

foraging for herbs and ⬜⬜⬜⬜⬜⬜⬜.

Three shapes of ⬜⬜⬜⬜⬜⬜ on a sassafras tree

Help me to ⬜⬜⬜⬜ where the yummy root beer leaves will be.

Elderberry, blackberry, and wild ⬜⬜⬜⬜⬜⬜⬜⬜⬜⬜ for jelly,

wild mint for when I have an ⬜⬜⬜⬜⬜ belly.

Wild green for a ⬜⬜⬜⬜⬜, mushrooms and ramps with our steak,

we give back to the ⬜⬜⬜⬜⬜ for the food that we take!

READING & COPYWORK

Copy the poem or story below and add color to the illustration.

_____

_____

_____

_____

_____

_____

_____

_____

_____

_____

_____

_____

_____

_____

_____

_____

_____

_____

_____

POETRY READING

Read the poem or story three times.

COME TAKE A WALK WITH ME

by Sarah Janisse Brown

Come, take a walk with me,
there is so much to see,
out past the wandering stream.
Come, and be still with me,
come, climb a hill with me,
out to the pastures of green.

Come, watch the clouds with me,
come, and be loud with me,
out where no others will hear.
All of our funny jokes,
inspired by silly goats,
smiling from ear to ear.

Come, let's climb in the trees,
come, search for hives and bees,
out where the wildflowers grow.
Come, take a walk with me,
there is so much to see-
follow the stream as is flows.

READING & COPYWORK

Copy the poem or story below and add color to the illustration.

_____
_____
_____
_____
_____
_____
_____
_____
_____
_____
_____
_____
_____
_____
_____
_____
_____
_____
_____
_____
_____
_____
_____
_____
_____
_____
_____

POETRY READING & WRITING

Read the poem or story and then fill in the missing words.

IN MY WINDOW

by MaKenzie Alberdi

Little plants in my window

Watching them grow,

sunlight shines through my window

nourishing them so.

Not much space in my window

yet still they grow,

little miracles in my window

I love them so.

IN MY WINDOW

Little ▢▢▢▢ in my window

▢▢▢▢▢ them grow,

sunlight ▢▢▢▢▢ through my window

▢▢▢▢▢▢▢ them so.

Not much ▢▢▢▢ in my window

yet ▢▢▢▢ they grow,

little ▢▢▢▢▢▢ in my window

I love ▢▢▢ so.

READING & COPYWORK

Copy the poem or story below and add color to the illustration.

_____

_____

_____

_____

_____

_____

_____

_____

_____

_____

_____

_____

_____

POETRY READING

Read the poem or story three times.

CHICKENS HERE & THERE

by Stefanie Risor Yates

Chickens sprinkled here and there,

in the field and on the chair,

basking in dust and perched high-

the large dotted one is so shy.

Singing their little egg-laying song,

in those mornings I hum along,

gathering each egg one by one,

getting the farm chores done.

On top of the coop roof, one stands,

carrying the small one in my hands.

On the patio and by the shed,

scratching in my garden bed,

following me as I walk along-

a small group behind me as I hum my song.

Chickens sprinkled one, two, three,

on one leg and under a tree.

Finished the farm chores; now I'll go.

In the distance the roosters crow.

Chickens sprinkled here and there;

my chickens are everywhere.

READING & COPYWORK

Copy the poem or story below and add color to the illustration.

_____
_____
_____
_____
_____
_____
_____
_____
_____
_____
_____
_____
_____
_____
_____
_____
_____
_____
_____

POETRY READING

Read the poem or story three times.

GOLDEN CRUST

By Julia Flock

Fold and fold and fold again,

hands covered in white dough.

Slide it in the glowing oven

and wait for bread to grow.

I press my nose up to the door

to watch the pale ball rise,

and soon the color starts to turn

before my very eyes.

The gold it turns is not the hue

that shines from heaps of treasure,

but that God turns the leaves in fall

to bring us truer pleasure.

READING & COPYWORK

Copy the poem or story below and add color to the illustration.

_____

_____

_____

_____

_____

_____

_____

_____

_____

_____

_____

_____

_____

_____

_____

_____

_____

_____

_____

_____

_____

_____

_____

_____

_____

_____

_____

_____

POETRY READING

Read the poem or story three times.

MAMA SHEEP

by Dekota Vaughn Bacon

Mama Sheep, so cozy and warm,

headed to bed, in a sweet little barn.

Two little lambs lay close by,

with their mama by their side.

Winter nights can be so cold,

but with Mama nearby,

they sleep sound in her wool.

The morning dew falls,

the warm sun shines.

Mama arises with lambs in line,

bouncing out to the field so green,

feeling the change- the first day of spring.

Mama Sheep keeps her lambs so near.

She's a good little mama;

her lambs have no fear.

READING & COPYWORK

Copy the poem or story below and add color to the illustration.

_____

_____

_____

_____

_____

_____

_____

_____

_____

_____

_____

_____

_____

_____

_____

_____

POETRY READING & WRITING

Read the poem or story and then fill in the missing words.

FLOCKS & HERDS

by Katie Keller Welte

At dawn we walk the pasture wide,

with feed and salt blocks by our side.

We check their coats, their stance, their eyes,

beneath the ever-changing skies.

We mend the fence, we watch the rain,

and note each limp or sign of strain.

Through seasons harsh and moments sweet,

their strength and trust make care complete.

FLOCKS & HERDS

At dawn we walk the ☐☐☐☐☐☐ wide,

with feed and salt ☐☐☐☐☐ by our side.

We check their coats, their ☐☐☐☐☐☐, their eyes,

beneath the ever-changing ☐☐☐☐.

We mend the ☐☐☐☐☐, we watch the rain,

and note each ☐☐☐☐ or sign of strain.

Through ☐☐☐☐☐☐☐ harsh and moments sweet,

their ☐☐☐☐☐☐ and trust make care complete.

READING & COPYWORK

Copy the poem or story below and add color to the illustration.

_____
_____
_____
_____
_____
_____
_____
_____
_____
_____
_____
_____
_____
_____
_____
_____

POETRY READING

Read the poem or story three times.

HOMEGROWN HERBS & VEGGIES

by Bear Sutcliffe, age 11

In April, when starting our seeds,

we think of all the herbs and veggies we'll need.

Seedlings grow and emerge till they're seen—

we check every day till we see bright green.

Dad fills the watering can, but it's my job to lug.

Pollinating bees are my favorite beneficial bug.

Lavender, basil, sage, and rosemary.

READING & COPYWORK

Copy the poem or story below and add color to the illustration.

_____
_____
_____
_____
_____
_____
_____
_____
_____
_____
_____
_____
_____
_____
_____
_____
_____

POETRY READING & WRITING

Read the poem or story and then fill in the missing words.

MICROGREENS

by Hannah Eroh

In trays of soil, seeds are sown,

by window's light, are quickly grown.

In days, not weeks, their leaves unfold,

a bounty of delights untold.

From kitchen sill to table's plate,

a burst of flavor they create.

Sprouts of blessing, green and slight,

a harvest of plenty with all their might.

MICROGREENS

In trays of soil, seeds are ☐☐☐☐☐,

by window's light, are ☐☐☐☐☐☐ grown.

In days, not weeks, their ☐☐☐☐☐☐ unfold,

a bounty of ☐☐☐☐☐☐☐ untold.

From ☐☐☐☐☐☐☐ sill to table's plate,

a burst of ☐☐☐☐☐☐ they create.

Sprouts of ☐☐☐☐☐☐☐☐, green and slight,

a harvest of ☐☐☐☐☐☐ with all their might.

READING & COPYWORK

Copy the poem or story below and add color to the illustration.

_____
_____
_____
_____
_____
_____
_____
_____
_____
_____
_____
_____
_____
_____
_____
_____
_____
_____
_____

POETRY READING

Read the poem or story three times.

BUILDING A GREENHOUSE

by Dale Smart-Felt

We gathered glass and timber with care,

for weeks we searched here, there, and everywhere.

We stacked it high in a tidy little heap,

waiting patiently to take the leap.

At last came the day to hammer and build–

A glasshouse dream that left us thrilled!

To grow warm plants from far-off lands,

with compost rich, it felt good in our hands.

We spread the soil, soft and deep,

and soon small bugs began to creep.

Now in our glassy, glowing dome,

the plants and bugs have found a home!

READING & COPYWORK

Copy the poem or story below and add color to the illustration.

_____
_____
_____
_____
_____
_____
_____
_____
_____
_____
_____
_____
_____
_____
_____
_____
_____
_____

POETRY READING

Read the poem or story three times.

MAPLE SYRUP

by Elijah Miller

Maple syrup—
we tap the tree.
Sap comes drip, drip, drop—
it just keeps flowing
When will it stop?
Collect and collect
until we have enough.
This is going to be delicious stuff!
Cooking and boiling
all day long—
the sweet smell is starting
to get strong.
The color is changing
from clear to brown.
The amount of liquid
is going down.
Pour it in jars—
Not a drip to waste!
It's finally time...
let's have a taste.

READING & COPYWORK

Copy the poem or story below and add color to the illustration.

_____
_____
_____
_____
_____
_____
_____
_____
_____
_____
_____
_____
_____
_____
_____
_____
_____
_____
_____
_____
_____
_____
_____
_____
_____
_____
_____
_____
_____

POETRY READING & WRITING

Read the poem or story and then fill in the missing words.

MY GARDEN

by Stefanie Risor Yates

This is my garden,

I'll plant it with care.

Soon I'll have vegetables

and herbs everywhere.

The sun will shine.

The rain will fall.

The seeds will sprout

and grow up tall.

MY GARDEN

This is my ⬚⬚⬚⬚⬚,

I'll ⬚⬚⬚⬚ it with care.

Soon I'll have ⬚⬚⬚⬚⬚⬚⬚⬚

and ⬚⬚⬚⬚ everywhere.

The sun will ⬚⬚⬚⬚⬚.

The ⬚⬚⬚⬚ will fall.

The seeds will ⬚⬚⬚⬚⬚⬚

and ⬚⬚⬚⬚ up tall.

READING & COPYWORK

Copy the poem or story below and add color to the illustration.

_____
_____
_____
_____
_____
_____
_____
_____
_____
_____
_____
_____

POETRY READING & WRITING

Read the poem or story and then fill in the missing words.

BUZZY BEE

by Desiree Lewis

Buzzy bee,
yellow fellow,
making honey,
never mellow.
Pollinating
high and low,
making all the
flowers grow.

BUZZY BEE

☐☐☐☐☐ bee,

yellow ☐☐☐☐☐,

☐☐☐☐☐ honey,

never ☐☐☐☐☐.

☐☐☐☐☐☐☐☐☐☐

high and ☐☐☐,

☐☐☐☐☐ all the

flowers ☐☐☐☐.

READING & COPYWORK

Copy the poem or story below and add color to the illustration.

_____

_____

_____

_____

_____

_____

_____

_____

_____

_____

_____

_____

_____

_____

POETRY READING

Read the poem or story three times.

NAMING GOATS

by Sarah Janisse Brown

My little sister named the goats when she was three years old.

My mother gave that job to her, and so the tale unfolds.

The first one's name it Smarfy-peps, the second Tootie-Lou.

The third goat's name is Tiny Tim - but he's the biggest in our zoo!

The twin kids' names are Bop and Pop, the white ones name is Blacky-Doo.

The brown one's name is Chocolate Soup, the speckled one is Sparky-Poo!

My little sister named our ducks when she was four or five.

She named them Giggles, Wiggles, Clucks, and named the duckling Apple Pie.

My little sister named the baby that our mama birthed last year.

I can't believe the name she picked! You would not believe your ears!

It wasn't James or Bob or Danny, like every normal kid.

She named him Sparky-Sharky-Ranny! He'll hate it when he's big!

I asked my mom why she would let my sister name our baby boy.

She said that it's her special gift - name-giving brings her joy!

READING & COPYWORK

Copy the poem or story below and add color to the illustration.

_____
_____
_____
_____
_____
_____
_____
_____
_____
_____
_____
_____
_____
_____
_____
_____
_____
_____
_____
_____
_____
_____
_____
_____
_____
_____
_____

POETRY READING & WRITING

Read the poem or story and then fill in the missing words.

RAISING SHEEP

by Dale Smart-Felt

Out in the paddock, woolly and loud,

our sheep strut 'round like they're quite proud.

They skip and nibble, baa and dash,

and sometimes bolt off in a flash.

We chase them 'round with all our might,

they dodge like pros in every sight.

But through the mess and muddy feet,

it's a woolly life – and they're the best!

RAISING SHEEP

Out in the ⬜⬜⬜⬜⬜⬜ , woolly and loud,

our sheep ⬜⬜⬜⬜⬜ 'round like they're quite proud.

They skip and ⬜⬜⬜⬜⬜⬜ , baa and dash,

and sometimes ⬜⬜⬜⬜ off in a flash.

We ⬜⬜⬜⬜⬜ them 'round with all our might,

they ⬜⬜⬜⬜⬜ like pros in every sight.

But ⬜⬜⬜⬜⬜⬜ the mess and muddy feet,

it's a ⬜⬜⬜⬜⬜⬜ – life and they're the best!

READING & COPYWORK

Copy the poem or story below and add color to the illustration.

_____

_____

_____

_____

_____

_____

_____

_____

_____

_____

_____

POETRY READING & WRITING

Read the poem or story and then fill in the missing words.

SMITTEN

By Sarah Janisse Brown

I traveled 'cross the nation, when I was just a little girl,

riding in the old white van, with plans to see the world!

From sea to shining sea we went, I can't forget the journey,

we started down in Florida, and headed for New Jersey.

And then we traveled way out west to see the redwoods tall,

into hills along the coast, we made another call.

We stopped to visit Auntie Sue, her homestead full of charm,

with horses, sheep, and hens galore — a working little farm.

She had fruit trees, flower beds, and kittens by the dozen.

I played with animals all-day, ignoring all my cousins!

I was only five years old, but deep inside I knew,

someday I'd have a little farm, and my own petting zoo.

I'd till the soil, plant the trees, and raise some baby lambs,

with garden beds and wooden sheds and homemade kitchen jams.

A horse, a goat, a pickup truck, and yes — a dozen kittens!

If you had asked about my dreams, you'd know that I was smitten!

READING & COPYWORK

Copy the poem or story below and add color to the illustration.

_____
_____
_____
_____
_____
_____
_____
_____
_____
_____
_____
_____
_____
_____
_____
_____
_____
_____
_____

POETRY READING

Read the poem or story three times.

SYRUP FOR MY PANCAKES

by Bear Sutcliffe, age 11

The nights are cold with flurries of snow,

but the days are warmer, making the sap flow.

We plot the maples on a map,

we drill a hole and set the tap.

As we sit by the fire and boil it down,

it feels like forever before turning golden brown.

The fresh syrup tastes good on my pancake –

it's so much better than the store-bought fake!

READING & COPYWORK

Copy the poem or story below and add color to the illustration.

_____
_____
_____
_____
_____
_____
_____
_____
_____
_____
_____
_____
_____
_____
_____
_____
_____
_____
_____
_____
_____

POETRY READING

Read the poem or story three times.

BAREFOOT

By Sarah Janisse Brown

Every time I go outside, and climb up in the hay,

I toss my shoes off as I glide, the slippy-sliddy way.

I run though grass so soft and green, and leave my shoes behind –

all is perfect, so it seems, long as I don't run blind.

Sometimes I skip through pastures too, underneath the sky so blue,

and wish I had been wearing shoes, every time I step in stinky poo!

BAREFOOT

Every time I go ☐☐☐☐☐, and ☐☐☐☐☐ up in the hay,

I toss my ☐☐☐☐☐ off as I ☐☐☐☐☐, the slippy-sliddy way.

I run though ☐☐☐☐☐ so soft and green, and leave my shoes ☐☐☐☐☐ –

all is ☐☐☐☐☐☐, so it seems, long as I don't run ☐☐☐☐.

Sometimes I skip through ☐☐☐☐☐☐☐ too, ☐☐☐☐☐☐☐☐☐ the sky so blue,

and wish I had been wearing ☐☐☐☐☐, every time I step in ☐☐☐☐☐ poo!

READING & COPYWORK

Copy the poem or story below and add color to the illustration.

_____

_____

_____

_____

_____

_____

_____

_____

_____

_____

_____

_____

POETRY READING

Read the poem or story three times.

TIME STANDS STILL

by Sarah Janisse Brown

The grandfather clock in the hall
makes no sounds - no, not at all.
No tick-tock,
no ringing bell
I stand still - stop -
with time to tell.
Always the same, I'm telling you,
just don't expect time-telling truth.
But even broken clocks, they say,
will tell the truth two times a day!
Trust this clock at seven, not nine -
seven o' clock at breakfast time,
seven 'o clock as we unwind.
That old clock is doing fine.
My mother says the clock must stay,
my father says it's in the way.
My grandma say's it once worked well,
'til Grandpa tried to fix the bell.
He said that it was just too loud,
and so he tried to take it out.
But when he did, the whole thing stopped-
My grandfather broke the clock.

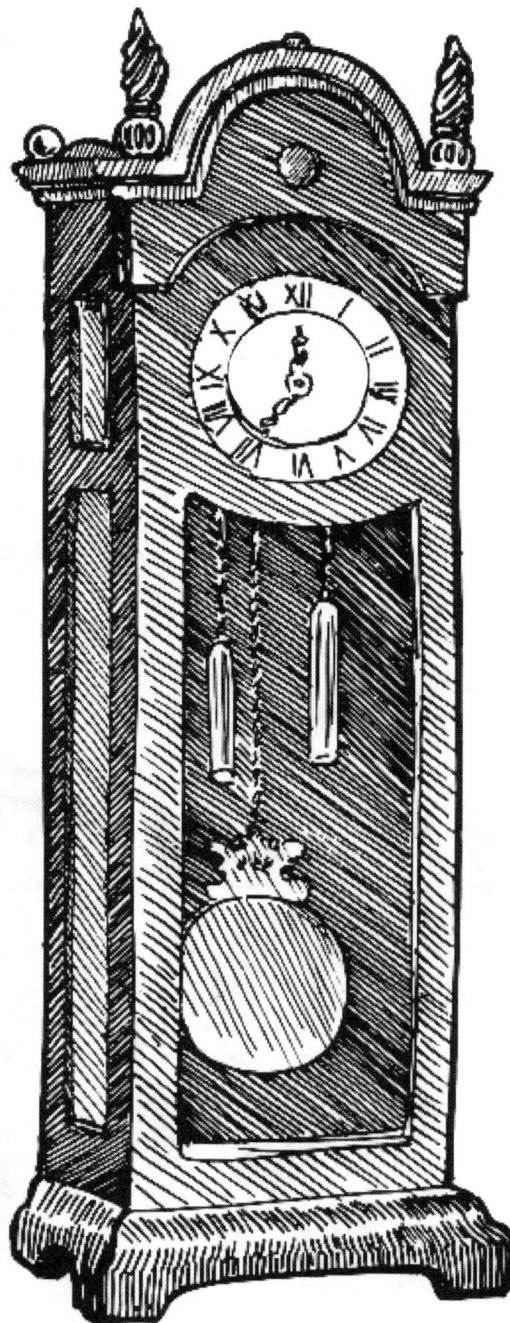

READING & COPYWORK

Copy the poem or story below and add color to the illustration.

POETRY READING & WRITING

Read the poem or story and then fill in the missing words.

RAISING CHICKENS

by Katie Keller Welte

Each morning starts with feed in hand,

fresh water poured, just as they planned.

The coop is cleaned, the bedding spread,

in straw-lined nests, the eggs are shed.

We watch for signs of health and cheer,

protect them when the hawks draw near.

Through winter's chill and summer's heat,

their trust and care make life complete.

RAISING CHICKENS

Each ☐☐☐☐☐☐☐ starts with feed in hand,

fresh ☐☐☐☐☐ poured, just as they planned.

The coop is ☐☐☐☐☐☐☐, the bedding spread,

in straw-lined ☐☐☐☐☐, the eggs are shed.

We ☐☐☐☐☐ for signs of health and cheer,

protect them when the ☐☐☐☐☐ draw near.

Through winter's ☐☐☐☐☐ and summer's heat,

their ☐☐☐☐☐ and care make life complete.

READING & COPYWORK

Copy the poem or story below and add color to the illustration.

_____

_____

_____

_____

_____

_____

_____

_____

_____

_____

_____

_____

_____

_____

_____

POETRY READING

Read the poem or story three times.

I FOUND MY GRANDMAS RECIPE

by Sarah Janisse Brown

I've got my grandma's recipe for *"Wild Garden Healing Tea."*

She said it's the one I need for tummy aches and frequent sneeze.

She said to drink it before bed; she said to heal my aching head.

She said to drink it when I wake – it's better than the pills some take.

It's full of flowers, leaves and seeds, and smells like spicy Christmas trees.

She adds some honey from her bees, and gives a lemon just one squeeze.

She said this is the fix I need, 'cause really, nothing wrong with me

that can't be fixed by sips of tea, when I trust grandma's recipe.

And when my friends and family sniffle, snuffle, cough, and sneeze

I'll share my grandma's recipe for *"Wild Garden Healing Tea."*

READING & COPYWORK

Copy the poem or story below and add color to the illustration.

_____

_____

_____

_____

_____

_____

_____

_____

_____

_____

_____

_____

_____

_____

_____

_____

_____

_____

_____

_____

_____

_____

_____

_____

POETRY READING

Read the poem or story three times.

MAKING FIREWOOD

By Dale Smart-Flett & Sarah Janisse Brown

On the edge of the forest, majestic and tall,

an old oak stood above it all.

It cleaned the air and gave us shade,

then fell to earth one stormy day.

My dad and I surveyed the scene-

brittle branches, no leaves of green.

I watched my father with his saw,

turning branches into logs.

To keep us warm when the cold winds roar,

we stacked the wood beside the door.

With gratitude for every tree-

for warmth, for shade, for air to breath.

READING & COPYWORK

Copy the poem or story below and add color to the illustration.

_____
_____
_____
_____
_____
_____
_____
_____
_____
_____
_____
_____
_____
_____
_____
_____
_____
_____
_____
_____
_____
_____
_____
_____
_____
_____
_____

POETRY READING

Read the poem or story three times.

RULES FOR FORAGING

by Annie Swihart McCosh

Before you taste those berries sweet,

be sure you know they're safe to eat.

Carefully consider how much you need-

leave some for others, and some for seed.

Is this plant rare or protected?

Then thoughts of picking should be rejected.

When you go out on a foraging mission,

don't harvest other's land without permission.

RULES FOR FORAGING

Before you taste those ☐☐☐☐☐☐☐ sweet,

be sure you know they're ☐☐☐☐ to eat.

Carefully ☐☐☐☐☐☐☐☐ how much you need-

☐☐☐☐☐ some for others and some for seed.

Is this plant rare or ☐☐☐☐☐☐☐☐?

Then thoughts of ☐☐☐☐☐☐ should be rejected.

When you go out on a ☐☐☐☐☐☐☐ mission,

don't harvest other's land without ☐☐☐☐☐☐☐☐☐☐.

READING & COPYWORK

Copy the poem or story below and add color to the illustration.

_____

_____

_____

_____

_____

_____

_____

_____

_____

_____

_____

_____

POETRY READING

Read the poem or story three times.

FLOWER GARDEN

By Julia Flock

Dirt stains run all up my arms
and streak across my face .
I laugh at my past dreams of gardens
as things of blooms and grace.
Nature is a messy thing-
that's why I love it so.
We need the dirt and worms and rain
to make the flowers grow.
But kitchen tables should be clean,
dinner scraps aside,
So I wash dirt from blooming stems
to bring God's world inside.

READING & COPYWORK

Copy the poem or story below and add color to the illustration.

_____
_____
_____
_____
_____
_____
_____
_____
_____
_____
_____
_____
_____
_____
_____
_____
_____
_____
_____
_____
_____
_____

POETRY READING

Read the poem or story three times.

THE WORM

By Ralph Bergengren

When the earth is turned in spring,

the worms are fat as anything.

and birds come flying all around

to eat the worms right off the ground.

They like the worms just as much as I—

like bread and milk and apple pie.

And once, when I was very young,

I put a worm right on my tongue.

I didn't like the taste a bit,

And so I didn't swallow it.

But oh, it makes my mother squirm

because she thinks I ate that worm!

READING & COPYWORK

Copy the poem or story below and add color to the illustration.

_____
_____
_____
_____
_____
_____
_____
_____
_____
_____
_____
_____
_____
_____
_____
_____

POETRY READING

Read the poem or story three times.

GATHERING WOOD

By Amber Hinkle

Gathering wood-

dusk is coming, and so is the night.

I can see the sun dipping low,

moving about with all our might,

to find the fuel to help our fire glow.

Now the teepee, cabin, star or lean-to -

nestle the kindling, nice and cozy.

Strike the match- one, maybe two-

see the tiny flame grow to rosy.

GATHERING WOOD

⬚⬚⬚⬚⬚⬚ wood-

dusk is coming, and so is the ⬚⬚⬚.

I can see the sun ⬚⬚⬚⬚ low,

moving ⬚⬚⬚ with all our might,

to find the ⬚⬚⬚ to help our fire glow.

Now the teepee, ⬚⬚⬚, star or lean-to -

nestle the ⬚⬚⬚⬚, nice and cozy.

Strike the ⬚⬚⬚- one, maybe two-

see the tiny ⬚⬚⬚ grow to rosy.

READING & COPYWORK

Copy the poem or story below and add color to the illustration.

_____

_____

_____

_____

_____

_____

_____

_____

_____

_____

_____

_____

_____

_____

POETRY READING

Read the poem or story three times.

MAKING BUTTER

By Amber Hinkle

The bread is rising with the sun.

Home is cozy with the scent of yeast.

I can already taste it.

The cream has risen, ready to be skimmed—

creamy, rich, sweet, liquid gold—

poured into the mixer to churn for some time.

The bread is in the oven.

I can already taste it.

Mixing and mixing, the liquid has firmed.

Curds form as buttermilk splashes.

Strain the buttermilk with a sip, maybe two.

The bread is cooling.

I can already taste it.

Rinse, rinse, rinse - salt, pat, mold.

Into the fridge, while the tidy commences.

Bread is sliced, toasted, and plated.

Fresh, creamy butter is spread on top.

I am ready to taste it.

READING & COPYWORK

Copy the poem or story below and add color to the illustration.

_____

_____

_____

_____

_____

_____

_____

_____

_____

_____

_____

_____

_____

_____

_____

_____

_____

_____

POETRY READING & WRITING

Read the poem or story and then fill in the missing words.

YUM YUM BREAD

by Stefanie Risor Yates

Yum yum bread,

so warm and fluffy.

Yum yum bread,

can't wait to have you in my tummy.

Yum yum bread,

bake faster, please!

Yum yum bread-

Yay! All done. that was a breeze!

YUM YUM BREAD

Yum yum ⬚⬚⬚⬚⬚,

so warm and ⬚⬚⬚⬚⬚.

Yum yum ⬚⬚⬚⬚⬚,

can't wait to have you in my ⬚⬚⬚⬚⬚.

Yum yum ⬚⬚⬚⬚⬚,

bake faster, ⬚⬚⬚⬚⬚⬚ !

Yum yum ⬚⬚⬚⬚⬚-

Yay! All done. That was a ⬚⬚⬚⬚⬚⬚ !

READING & COPYWORK

Copy the poem or story below and add color to the illustration.

_____

_____

_____

_____

_____

_____

_____

_____

_____

_____

_____

_____

_____

_____

12 MONTHS OF HOMESTEADING POETRY

Think of all the things that happen on a homestead year round! Take some time to organize Homesteading Activities by the season, and get ready to write and illustrate twelve poems!

SPRING

1. _____

2. _____

3. _____

4. _____

5. _____

SUMMER

1. _____
2. _____
3. _____
4. _____
5. _____

AUTUMN

1. _____
2. _____
3. _____
4. _____
5. _____

WINTER

1. _____
2. _____
3. _____
4. _____
5. _____

JANUARY

FEBRUARY

_____

_____

_____

_____

_____

_____

_____

_____

_____

_____

_____

_____

_____

_____

MARCH

_____
_____
_____
_____
_____
_____
_____
_____
_____
_____
_____
_____
_____
_____
_____

APRIL

MAY

_____

_____

_____

_____

_____

_____

_____

_____

_____

_____

_____

_____

_____

_____

_____

JUNE

_____

_____

_____

_____

_____

_____

_____

_____

_____

_____

_____

_____

_____

_____

_____

JULY

_____

_____

_____

_____

_____

_____

_____

_____

_____

_____

_____

_____

_____

_____

_____

AUGUST

_____

_____

_____

_____

_____

_____

_____

_____

_____

_____

_____

_____

_____

_____

_____

SEPTEMBER

_____
_____
_____
_____
_____
_____
_____
_____
_____
_____
_____
_____
_____
_____
_____

OCTOBER

_____

_____

_____

_____

_____

_____

_____

_____

_____

_____

_____

_____

_____

_____

_____

_____

NOVEMBER

_____
_____
_____
_____
_____
_____
_____
_____
_____
_____
_____
_____
_____
_____
_____

DECEMBER

_____

_____

_____

_____

_____

_____

_____

_____

_____

_____

_____

_____

_____

_____

_____

_____

THE Thinking TREE

PUBLISHING COMPANY

Sarah Janisse Brown

www.ingramcontent.com/pod-product-compliance
Lightning Source LLC
Chambersburg PA
CBHW072000220326
41599CB00034BA/7063